Livonia Public Library
CARL SANDBURG BRANCH #21
30100 W. 7 Mile Road
Livonia, Mich. 48152
248-893-4010

JOBS KIDS WANT

WHAT'S IT REALLY LIKE TO BE A
MECHANIC?

Carl Sandburg Library
30100 W. 7 Mile Rd.
Livonia, MI 48152-1918
(248) 893-4010

CHRISTINE HONDERS

PowerKiDS
press

New York

Published in 2020 by The Rosen Publishing Group, Inc.
29 East 21st Street, New York, NY 10010

Copyright © 2020 by The Rosen Publishing Group, Inc.

All rights reserved. No part of this book may be reproduced in any form without permission in writing from the publisher, except by a reviewer.

First Edition

Editor: Greg Roza
Book Design: Michael Flynn

Photo Credits: Cover, p. 1 Westend61/Getty Images; pp. 4, 6, 8, 10, 12, 14, 16, 18, 20, 22 (background) Apostrophe/Shutterstock.com; p. 5 Mint Images/Getty Images; p. 7 FrameStockFootages/Shutterstock.com; pp. 9 (mechanic), 13 Lemusique/Shutterstock.com; p. 9 (engine) Pro3DArtt/Shutterstock.com; p. 11 welcomia/ Shutterstock.com; p. 15 bondvit/Shutterstock.com; p. 17 Peathegee Inc/Blend Images/Getty Images; p. 19 goodluz/Shutterstock.com; p. 21 SeventyFour/Shutterstock.com; p. 22 simm49/Shutterstock.com.

Cataloging-in-Publication Data

Names: Honders, Christine.
Title: What's it really like to be a mechanic? / Christine Honders.
Description: New York : PowerKids Press, 2020. | Series: Jobs kids want | Includes glossary and index.
Identifiers: ISBN 9781538349922 (pbk.) | ISBN 9781538349946 (library bound) | ISBN 9781538349939 (6pack)
Subjects: LCSH: Motor vehicles–Maintenance and repair–Vocational guidance–Juvenile literature. | Mechanics (Persons)–Juvenile literature.
Classification: LCC TL152.H66 2020 | DDC 629.28'7023–dc23
Manufactured in the United States of America

CPSIA Compliance Information: Batch #CSPK19. For Further Information contact Rosen Publishing, New York, New York at 1-800-237-9932.

CONTENTS

Who's Going to Fix the Car? 4

Masters of Machines 6

What's a Motor? 8

Auto Mechanics 10

Tools of the Job. 12

In the Shop 14

Long, Hard Days 16

Becoming a Mechanic 18

Other Important Skills. 20

A Love for Machines. 22

Glossary 23

Index 24

Websites 24

Who's Going to Fix the Car?

It's probably happened to you. You're ready to go somewhere and your parents say the car won't start! If they can't start it, what do they do? They go to a mechanic. Mechanics can fix cars so they run as good as new.

Masters of Machines

Mechanics are masters of **machines**. They use special tools to build and fix them. They know how to take different machines apart and figure out what's broken. Then, they fix the broken part and put it back together.

What's a Motor?

A motor is a machine that turns fuel into **energy** to work. A car motor, or engine, uses gasoline as fuel. The engine makes little **explosions** that turn the gasoline into energy! That energy is what makes the car run.

car engine

Auto Mechanics

Most mechanics are auto mechanics. They fix all kinds of **vehicles**. They also **inspect** them to make sure they're running safely. Some mechanics have training in fixing certain parts of vehicles, such as the brakes. Some only work on motorcycles, or tractor-trailer trucks.

Tools of the Job

Mechanics use hand tools, like screwdrivers and wrenches. They also use special tools that work using **air pressure**. They can tighten bolts stronger and faster than any human! Today's cars have computers built into them. Mechanics use electronic tools to check the computers for problems.

13

In the Shop

A mechanic's shop must have bright lights and **ventilation**. Mechanics work in open areas called pits. Pits can be very hot in the summer and freezing cold in the winter. Mechanics work with greasy engines all day. There's no way to stay clean!

Long, Hard Days

Mechanics work long hours. They often work nights and weekends. They can also get hurt on the job. They work with hot engines and sharp tools. They get burns, cuts, and bruises on their hands. They get sore from lifting heavy vehicle parts.

Becoming a Mechanic

Mechanics need a high school degree. Taking computer and electronics classes is a plus. Then, they must complete a training program in auto mechanics. Most mechanics learn on the job for another year before taking a test to become a **certified** mechanic.

19

Other Important Skills

Mechanics are good with their hands. They repair small parts. They're also good at solving problems. They pay attention to details to figure out what's broken. Mechanics are good with people. They listen carefully. Then, they explain what's wrong and how they're going to fix it.

A Love for Machines

If you love machines, and don't mind getting dirty, you might want to be a mechanic. It's not just a great job, it's a great skill to learn. Then, if your car won't start, you can fix it yourself!

GLOSSARY

air pressure: The force created by air when it is squeezed into a small container.

certified: Has special training in a type of work.

energy: The power needed to work or act.

explosion: A sudden, noisy, and hot bursting out of something.

inspect: To look at closely.

machine: A device with moving parts that does some kind of work when it is provided with power.

vehicle: Something used to move people or objects.

ventilation: A way of providing fresh air.

INDEX

C
car, 4, 8, 12, 22
computer, 12, 18

E
engine, 8, 14, 16

F
fix, 4, 6, 10, 20, 22

M
machine, 6, 8, 22
motor, 8
motorcycle, 10

P
pit, 14

S
screwdriver, 12

T
tool, 6, 12, 16
training, 10, 18

W
wrench, 12

WEBSITES

Due to the changing nature of Internet links, PowerKids Press has developed an online list of websites related to the subject of this book. This site is updated regularly. Please use this link to access the list: www.powerkidslinks.com/JKW/mechanic